Louis Agassiz as a Teacher; Illustrative Extracts on his Method of Instruction

LOUIS AGASSIZ

LOUIS AGASSIZ AS A TEACHER

ILLUSTRATIVE EXTRACTS ON
HIS METHOD OF INSTRUCTION

WITH AN INTRODUCTORY NOTE

BY

LANE COOPER

PROFESSOR OF THE ENGLISH LANGUAGE AND LITERATURE
IN CORNELL UNIVERSITY

THE COMSTOCK PUBLISHING CO.
ITHACA, NEW YORK
1917

Open access edition funded by the National Endowment for the
Humanities/Andrew W. Mellon Foundation Humanities Open Book
Program.

ISBN 978-1-5017-4056-5 (pbk.: alk. paper)
ISBN 978-1-5017-4058-9 (pdf)
ISBN 978-1-5017-4057-2 (epub/mobi)

Librarians: A CIP catalog record for this book is available from the
Library of Congress

PREFACE

IF it be asked why a teacher of English should be moved to issue this book on Agassiz, my reply might be: 'Read the Introductory Note'—for the answer is there. But doubtless the primary reason is that I have been taught, and I try to teach others, after a method in essence identical with that employed by the great naturalist. And I might go on to show in some detail that a doctoral investigation in the humanities, when the subject is well chosen, serves the same purpose in the education of a student of language and literature as the independent, intensive study of a living or a fossil animal, when prescribed by Agassiz to a beginner in natural science. But there is no need to elaborate the point. Of those who are likely to examine the book, some already know the underlying truth involved, others will grasp it when it is first presented to them (and for these my slight and pleasant labors are designed), and the rest will

[v]

find a stumbling-block and foolishness—save for the entertainment to be had in the reading of biography.

I have naturally kept in mind the needs of my own students, past and present, yet I believe these pages may be useful to students of natural science as well as to those who concern themselves with the humanities. We live in an age of narrow specialization—at all events in America. · Agassiz was a specialist, but not a 'narrow' one. His example should therefore be salutary to those persons, on the one hand, who think that a man can have general culture without knowing some one thing from the bottom up, and, on the other, to those who immerse themselves and their pupils blindly in special investigation, without thought of the *prima philosophia* that gives life and meaning to all particular knowledge. There can be no doubt that science and scholarship in this country are suffering from a lack of sympathy and contact between the devotees of the several branches, and for want of definite efforts to bridge the gaps between various disciplines wherever this is possible. It may not often

PREFACE

be possible until men of science generally again take up the study of Plato and Aristotle, or at least busy themselves, as did Agassiz, with some comprehensive modern philosopher like Schelling. · But it should not be very hard for those who are engaged in the biological sciences and those who are given to literary pursuits to realize that they are alike interested in the manifestations of one and the same thing, the principle of life. · In Agassiz himself the vitality of his studies and the vitality of the man are easily identified. ·

In conclusion I must thank the publishers, Houghton Mifflin Company, for the use of selections from the copyright books of Mrs. Agassiz and Professor Shaler; these and all other obligations are, I trust, indicated in the proper places by footnotes. I owe a special debt of gratitude to Professor Burt G. Wilder for his interest and help throughout.

LANE COOPER

CORNELL UNIVERSITY,
April 7, 1917.

CONTENTS

I

INTRODUCTORY NOTE

WHEN the question was put to Agassiz, 'What do you regard as your greatest work?' he replied: 'I have taught men to observe.' . And in the preamble to his will he described himself in three words as 'Louis Agassiz, Teacher.'

We have more than one reason to be interested in the form of instruction employed by so eminent a scientist as Agassiz. In the first place, it is much to be desired that those who concern themselves with pedagogy should give relatively less heed to the way in which subjects, abstractly considered, ought to be taught, and should pay more attention than I fear has been paid to the way in which great and successful teachers actually have taught their pupils. As in other fields of human endeavor, so in teaching: there is a portion of the art that cannot be taken over by one person from another, but there is a portion, and a larger

[1]

one than at first sight may appear, that can be so taken over, and can be almost directly utilized. Nor is the possible utility of imitation diminished, but rather increased, when we contemplate the method of a teacher like Agassiz, whose mental operations had the simplicity of genius, and in whose habits of instruction the fundamentals of a right procedure become very obvious. ·

Yet there is a second main reason for our interest. Within recent years we have witnessed an extraordinary development in certain studies, which, though superficially different from those pursued by Agassiz, have an underlying bond of unity with them, but which are generally carried on without reference to principles governing the investigation of every organism and all organic life. I have in mind, particularly, the spread of literary and linguistic study in America during the last few decades, and the lack of a common standard of judgment among those who engage in such study. Most persons do not, in fact, discern the close, though not obvious, relation between investigation in biology or zoology and the observation and comparison of those organic forms which we call forms of literature and works of art. ˙Yet the notion that a poem or a speech should possess the organic structure, as it were, of a living creature is basic in the thought of the great literary critics of all time. ˙So Aristotle, a zoologist as well as a systematic student of literature, compares the essential structure of a tragedy to the form of an animal. ˙And so Plato, in the *Phaedrus*, makes Socrates say: 'At any rate, you will allow that every discourse ought to be a living creature, having a body of its own, and a head and feet; there should be a middle, beginning, and end, adapted to one another and to the whole.' ˙ It would seem that to Plato an oration represents an organic idea in the mind of the human creator, the orator, just as a living animal represents a constructive idea in the mind of God. Now it happens that Agassiz, considered in his philosophical relations, was a Platonist, since he clearly believed that the forms of nature expressed the eternal ideas of a divine intelligence.

Accordingly, his method of teaching cannot fail to be illuminating to the teacher of literature—or to the teacher of language, either, since· each language as a whole, and also the component parts of language, words, for instance, are living and growing forms, and must be studied as organisms. · We have perhaps heard too much of 'laboratory' methods in the teaching of English and the like; but none of us has heard too much about the fundamental operations of observation and comparison in the study of living forms, or of the way in which great teachers have developed the original powers of the student. It is simply the fact that, reduced to the simplest terms, there is but a single method of investigating the objects of natural science and the productions of human genius. · We study a poem, the work of man's art, in the same way that Agassiz made Shaler study a fish, the work of God's art; the object in either case is to discover the relation between form or structure and function or essential effect. · It was no chance utterance of Agassiz when he said that a year or two of natural history, studied as he understood it, would give the best kind of training for any other sort of mental work.

[4]

INTRODUCTORY NOTE

The following passages will illustrate Agassiz's ideals and practice in teaching, the emphasis being laid upon his dealings with special students. A few biographical details are introduced in order to round out our conception of the personality of the teacher himself. Toward the close, certain of his opinions are given in his own words.

I would call special attention to an extract from Boeckh's *Encyclopädie*, and another from the *Symposium* of Plato, on pp. 69–74, and to the similarity between the method of study there enjoined upon the student of the humanities, or indeed of all art and nature, and the method imposed by Agassiz upon the would-be entomologist who was compelled first of all to observe a fish. · In reforming the mind it is well to begin by contemplating some structure we never have seen before, concerning which we have no, or the fewest possible, preconceptions. ·

II

AGASSIZ AT NEUCHÂTEL[1]

[IN the autumn of the year 1832] Agassiz assumed the duties of his professorship at Neuchâtel. His opening lecture, upon the relations between the different branches of natural history and the then prevailing tendencies of all the sciences, was given on the 12th of November . . . at the Hotel de Ville. Judged by the impression made, upon the listeners as recorded at the time, this introductory discourse must have been characterized by the same broad spirit of generalization which marked Agassiz's later teaching. · Facts in his hands fell into their orderly relation as parts of a connected whole, and were never presented merely as special or isolated phenomena. · From the beginning his success as an instructor was undoubted. He had, indeed, now entered upon the occupation which was to be from youth to

[1] From E. C. Agassiz, *Louis Agassiz, his Life and Correspondence*, pp. 206 ff. Boston, Houghton Mifflin Company, 1885.

[6]

old age the delight of his life. Teaching was a passion with him, and his power over his pupils might be measured by his own enthusiasm. He was intellectually, as well as socially, a democrat, in the best sense. He delighted to scatter broadcast the highest results of thought and research, and to adapt them even to the youngest and most uninformed minds. In his later American travels he would talk of glacial phenomena to the driver of a country stagecoach among the mountains, or to some workman, splitting rock at the road-side, with as much earnestness as if he had been discussing problems with a brother geologist; he would take the common fisherman into his scientific confidence, telling him the intimate secrets of fish-structure or fish-embryology, till the man in his turn became enthusiastic, and began to pour out information from the stores of his own rough and untaught habits of observation. Agassiz's general faith in the susceptibility of the popular intelligence, however untrained, to the highest truths of nature, was contagious, and he created or developed that in which he believed. . . .

[7]

Beside his classes at the gymnasium, Agassiz collected about him, by invitation, a small audience of friends and neighbors, to whom he lectured during the winter on botany, on zoology, on the philosophy of nature. The instruction was of the most familiar and informal character, and was continued in later years for his own children and the children of his friends. In the latter case the subjects were chiefly geology and geography in connection with botany, and in favorable weather the lessons were usually given in the open air. . . . From some high ground affording a wide panoramic view Agassiz would explain to them the formation of lakes, islands, rivers, springs, water-sheds, hills, and valleys. . . .

When it was impossible to give the lessons out of doors, the children were gathered around a large table, where each one had before him or her the specimens of the day, sometimes stones and fossils, sometimes flowers, fruits, or dried plants. To each child in succession was explained separately what had first been told to all collectively. . . . The children took their own share in the instruction, and were them-

[8]

selves made to point out and describe that which had just been explained to them. They took home their collections, and as a preparation for the next lesson were often called upon to classify and describe some unusual specimen by their own unaided efforts.

III

AGASSIZ AT HARVARD[1]

O N his return to Cambridge at the end of
September [1859], Agassiz found the
Museum building well advanced. It
was completed in the course of the next year,
and the dedication took place on the 13th of
November, 1860. The transfer of the collections to their new and safe abode was made as
rapidly as possible, and the work of developing
the institution under these more favorable
conditions moved steadily on. The lecture-
rooms were at once opened, not only to students,
but to other persons not connected with the
University. Especially welcome were teachers
of schools, for whom admittance was free. It
was a great pleasure to Agassiz thus to renew
and strengthen his connection with the teachers
of the State, with whom, from the time of his
arrival in this country, he had held most cordial
relations, attending the Teachers' Institutes,

[1] From E. C. Agassiz, *Louis Agassiz, his Life and Correspondence*, pp. 564 ff. Boston, Houghton Mifflin Company,
1885.

[10]

visiting the normal schools, and associating himself actively, as far as he could, with the interests of public education in Massachusetts. From this time forward his college lectures were open to women as well as to men. He had great sympathy with the desire of women for larger and more various fields of study and work, and a certain number of women have always been employed as assistants at the Museum. The story of the next three years was one of unceasing but seemingly uneventful work. The daylight hours from nine or ten o'clock in the morning were spent, with the exception of the hour devoted to the school, at the Museum, not only in personal researches and in lecturing, but in organizing, distributing, and superintending the work of the laboratories, all of which was directed by him. Passing from bench to bench, from table to table, with a suggestion here, a kindly but scrutinizing glance there, he made his sympathetic presence felt by the whole establishment. No man ever exercised a more genial personal influence over his students and assistants.

His initiatory steps in teaching special

2

students of natural history were not a little discouraging. Observation and comparison being in his opinion the intellectual tools most indispensable to the naturalist, his first lesson was one in *looking.* He gave no assistance; he simply left his student with the specimen, telling him to use his eyes diligently, and report upon what he saw. He returned from time to time to inquire after the beginner's progress, but he never asked him a leading question, never pointed out a single feature of the structure, never prompted an inference or a conclusion. This process lasted sometimes for days, the professor requiring the pupil not only to distinguish the various parts of the animal, but to detect also the relation of these details to more general typical features. His students still retain amusing reminiscences of their despair when thus confronted with their single specimen; no aid to be had from outside until they had wrung from it the secret of its structure. But all of them have recognized the fact that this one lesson in looking, which forced them to such careful scrutiny of the object before them, influenced all their subsequent habits

of observation, whatever field they might choose
for their special subject of study. . . . .
But if Agassiz, in order to develop inde-
pendence and accuracy of observation, threw
his students on their own resources at first,
there was never a more generous teacher in the
end than he. All his intellectual capital was
thrown open to his pupils. His original ma-
terial, his unpublished investigations, his most
precious specimens, his drawings and illustra-
tions were at their command. This liberality
led in itself to a serviceable training, for he
taught them to use with respect the valuable,
often unique, objects entrusted to their care.
Out of the intellectual good-fellowship which
he established and encouraged in the laboratory
grew the warmest relations between his students
and himself. Many of them were deeply at-
tached to him, and he was extremely dependent
upon their sympathy and affection. By some
among them he will never be forgotten. He is
still their teacher and their friend, scarcely
more absent from their work now than when
the glow of his enthusiasm made itself felt in
his personal presence.

IV

HOW AGASSIZ TAUGHT PROFESSOR SHALER[1]

AT the time of my secession from the
humanities, Agassiz was in Europe; he
did not return, I think, until the autumn
of 1859. I had, however, picked up several
acquaintances among his pupils, learned what
they were about, and gained some notion of
his methods. After about a month he returned,
and I had my first contact with the man who
was to have the most influence on my life of
any of the teachers to whom I am indebted.
I shall never forget even the lesser incidents
of this meeting, for this great master by his
presence gave an importance to his surround-
ings, so that the room where you met him, and
the furniture, stayed with the memory of him.

· When I first met Louis Agassiz, he was still
in the prime of his admirable manhood; though

[1] From *The Autobiography of Nathaniel Southgate Shaler*,
pp. 93–100. Boston, Houghton Mifflin Company, 1907.

[14]

he was then fifty-two years old, and had passed his constructive period, he still had the look of a young man. His face was the most genial and engaging that I had ever seen, and his manner captivated me altogether. But as I had been among men who had a free swing, and for a year among people who seemed to me to be cold and super-rational, hungry as I doubtless was for human sympathy, Agassiz's welcome went to my heart—I was at once his captive. · It has been my good chance to see many men of engaging presence and ways, but I have never known his equal.

As the personal quality of Agassiz was the greatest of his powers, and as my life was greatly influenced by my immediate and enduring affection for him, I am tempted to set forth some incidents which show that my swift devotion to my new-found master was not due to the accidents of the situation, or to any boyish fancy. I will content myself with one of those stories, which will of itself show how easily he captivated men, even those of the ruder sort. Some years after we came together, when indeed I was formally his assistant,—

I be'ieve it was in 1866,—he became much interested in the task of comparing the skeletons of thoroughbred horses with those of common stock. I had at his request tried, but without success, to obtain the bones of certain famous stallions from my acquaintances among the racing men in Kentucky. Early one morning there was a fire, supposed to be incendiary, in the stables in the Beacon Park track, a mile from the College, in which a number of horses had been killed, and many badly scorched. I had just returned from the place, where I had left a mob of irate owners and jockeys in a violent state of mind, intent on finding some one to hang. I had seen the chance of getting a valuable lot of stallions for the Museum, but it was evident that the time was most inopportune for suggesting such a disposition of the remains. Had I done so, the results would have been, to say the least, unpleasant.

As I came away from the profane lot of horsemen gathered about the ruins of their fortunes or their hopes, I met Agassiz almost running to seize the chance of specimens. I told him to come back with me, that we must wait until the mob had spent its rage; but he kept on. I told him further that he risked spoiling his good chance, and finally that he would have his head punched; but he trotted on. I went with him, in the hope that I might protect him from the consequences of his curiosity. When we reached the spot, there came about a marvel; in a moment he had all those raging men at his command. He went at once to work with the horses which had been hurt, but were savable. His intense sympathy with the creatures, his knowledge of the remedies to be applied, his immediate appropriation of the whole situation, of which he was at once the master, made those rude folk at once his friends. Nobody asked who he was, for the good reason that he was heart and soul of them. When the task of helping was done, then Agassiz skilfully came to the point of his business—the skeletons—and this so dexterously and sympathetically, that the men were, it seemed, ready to turn over the living as well as the dead beasts for his service. I have seen a lot of human doing, much of it critically as actor or near observer, but this was in many ways the greatest. The supreme art of it was in the use of a perfectly spontaneous and most actually sympathetic motive to gain an end.' With others, this state of mind would lead to affectation; with him, it in no wise diminished the quality of the emotion. He could measure the value of the motive, but do it without lessening its moral import.

As my account of Agassiz's quality should rest upon my experiences with him, I shall now go on to tell how and to what effect he trained me. In that day there were no written examinations on any subjects to which candidates for the Lawrence Scientific School had to pass. The professors in charge of the several departments questioned the candidates, and determined their fitness to pursue the course of study they desired to undertake. . Few or none who had any semblance of an education were denied admission to Agassiz's laboratory. At that time, the instructors had, in addition to their meagre salaries—his was then $2,500 per annum,—the regular fees paid in by the students under his charge. So I was promptly assured that I was admitted. . Be it said, how-

ever, that he did give me an effective oral
examination, which, as he told me, was intended to show whether I could expect to go
forward to a degree at the end of four years
of study. On this matter of the degree he was
obdurate, refusing to recommend some who
had been with him for many years, and had
succeeded in their special work, giving as
reason for his denial that they were 'too
ignorant.'

The examination Agassiz gave me was
directed first to find that I knew enough Latin
and Greek to make use of those languages;
that I could patter a little of them evidently
pleased him. He didn't care for those detest-
able rules for scanning. Then came German
and French, which were also approved: I could
read both, and spoke the former fairly well.
He did not probe me in my weakest place,
mathematics, for the good reason that, badly
as I was off in that subject, he was in a worse
plight. Then asking me concerning my reading, he found that I had read the *Essay on
Classification,* and had noted in it the influence
of Schelling's views. Most of his questioning

related to this field, and the more than fair beginning of our relations then made was due to the fact that I had some enlargement on that side. So, too, he was pleased to find that I had managed a lot of Latin, Greek, and German poetry, and had been trained with the sword. He completed this inquiry by requiring that I bring my foils and masks for a bout. In this test he did not fare well, for, though not untrained, he evidently knew more of the *Schläger* than of the rapier. He was heavy-handed, and lacked finesse. This, with my previous experience, led me to the conclusion that I had struck upon a kind of tutor in Cambridge not known in Kentucky.

While Agassiz questioned me carefully as to what I had read and what I had seen, he seemed in this preliminary going over in no wise concerned to find what I knew about fossils, rocks, animals, and plants; he put aside the offerings of my scanty lore. This offended me a bit, as I recall, for the reason that I thought I knew, and for a self-taught lad really did know, a good deal about such matters, especially as to the habits of insects, particularly spiders.

It seemed hard to be denied the chance to make my parade; but I afterward saw what this meant—that he did not intend to let me begin my tasks by posing as a naturalist. The beginning was indeed quite different, and, as will be seen, in a manner that quickly evaporated my conceit. It was made and continued in a way I will now recount.

. Agassiz's laboratory was then in a rather small two-storied building, looking much like a square dwelling-house, which stood where the College Gymnasium now stands. . . . Agassiz had recently moved into it from a shed on the marsh near Brighton bridge, the original tenants, the engineers, having come to riches in the shape of the brick structure now known as the Lawrence Building. In this primitive establishment Agassiz's laboratory, as distinguished from the storerooms where the collections were crammed, occupied one room about thirty feet long and fifteen feet wide— what is now the west room on the lower floor of the edifice. In this place, already packed, I had assigned to me a small pine table with a rusty tin pan upon it. . . . .

When I sat me down before my tin pan,
Agassiz brought me a small fish, placing it
before me with the rather stern requirement
that I should study it, but should on no account
talk to any one concerning it, nor read anything
relating to fishes, until I had his permission
so to do. To my inquiry, 'What shall I do?'
he said in effect: 'Find out what you can without damaging the specimen; when I think that
you have done the work I will question you.'
In the course of an hour I thought I had
compassed that fish; it was rather an unsavory
object, giving forth the stench of old alcohol,
then loathsome to me, though in time I came to
like it. Many of the scales were loosened so
that they fell off. It appeared to me to be a
case for a summary report, which I was anxious
to make and get on to the next stage of the
business. But Agassiz, though always within
call, concerned himself no further with me that
day, nor the next, nor for a week. At first,
this neglect was distressing; but I saw that it
was a game, for he was, as I discerned rather
than saw, covertly watching me. So I set
my wits to work upon the thing, and in the

course of a hundred hours or so thought I had done much—a hundred times as much as seemed possible at the start. I got interested in finding out how the scales went in series, their shape, the form and placement of the teeth, etc. Finally, I felt full of the subject, and probably expressed it in my bearing; as for words about it then, there were none from my master except his cheery 'Good morning.' At length, on the seventh day, came the question, 'Well?' and my disgorge of learning to him as he sat on the edge of my table puffing his cigar. At the end of the hour's telling, he swung off and away, saying: 'That is not right.' Here I began to think that, after all, perhaps the rules for scanning Latin verse were not the worst infliction in the world. Moreover, it was clear that he was playing a game with me to find if I were capable of doing hard, continuous work without the support of a teacher, and this stimulated me to labor. I went at the task anew, discarded my first notes, and in another week of ten hours a day labor I had results which astonished myself and satisfied him. Still there was no trace of praise in

[23]

words or manner. He signified that it would do by placing before me about a half a peck of bones, telling me to see what I could make of them, with no further directions to guide me. I soon found that they were the skeletons of half a dozen fishes of different species; the jaws told me so much at a first inspection. The task evidently was to fit the separate bones together in their proper order. Two months or more went to this task with no other help than an occasional looking over my grouping with the stereotyped remark: 'That is not right.' Finally, the task was done, and I was again set upon alcoholic specimens—this time a remarkable lot of specimens representing, perhaps, twenty species of the side-swimmers or Pleuronectidae.

I shall never forget the sense of power in dealing with things which I felt in beginning the more extended work on a group of animals. I had learned the art of comparing objects, which is the basis of the naturalist's work. At this stage I was allowed to read, and to discuss my work with others about me. I did both eagerly, and acquired a considerable knowledge

[24]

of the literature of ichthyology, becoming especially interested in the system of classification, then most imperfect. I tried to follow Agassiz's scheme of division into the order of ctenoids and ganoids, with the result that I found one of my species of side-swimmers had cycloid scales on one side and ctenoid on the other. This not only shocked my sense of the value of classification in a way that permitted of no full recovery of my original respect for the process, but for a time shook my confidence in my master's knowledge. At the same time I had a malicious pleasure in exhibiting my 'find' to him, expecting to repay in part the humiliation which he had evidently tried to inflict on my conceit. To my question as to how the nondescript should be classified he said: 'My boy, there are now two of us who know that.' ·

This incident of the fish made an end of my novitiate. After that, with a suddenness of transition which puzzled me, Agassiz became very communicative; we passed indeed into the relation of friends of like age and purpose, and he actually consulted me. as to what I should like to take up as a field of study. Finding that I wished to devote myself to geology, he set me to work on the Brachiopoda as the best group of fossils to serve as data in determining the Palaeozoic horizons. So far as his rather limited knowledge of the matter went, he guided me in the field about Cambridge, in my reading, and to acquaintances of his who were concerned with earth structures. I came thus to know Charles T. Jackson, Jules Marcou, and, later, the brothers Rogers, Henry and James. At the same time I kept up the study of zoology, undertaking to make myself acquainted with living organic forms as a basis for a knowledge of fossils.

V

HOW AGASSIZ TAUGHT PROFESSOR VERRILL[1]

IN regard to the methods of instruction of Agassiz I must say that so far as I saw and experienced he had no regular or fixed method, except that his plan was to make young students depend on natural objects rather than on statements in books. · To that end he treated each one of his new students differently, according to the amount of knowledge and experience that the student had previously acquired, and often in line with what the student had done before. Not infrequently young men came to him who were utterly destitute of any knowledge or ability to study natural science, or zoology in particular, but had an idea that it would be a 'soft snap,' as the boys say. In such cases he often did give them a lot of mixed stuff to mull over, to see

[1] From a private letter from Professor Addison Emery Verrill to Lane Cooper. The extract is printed with the consent of Professor Verrill.

3

what they could do, and also to discourage those that seemed unfit. Sometimes he was mistaken, of course, and the student would persevere and stay on—and sometimes turned out well later. In fact, his treatment was highly and essentially individualistic.

In my own case, he questioned me closely as to what I had previously done and learned. He found I had made collections of birds, mammals, plants, etc., and had mounted and identified them for several years, and in that way was not a beginner exactly. I remember that before I had been with him six months he told me I knew more zoology than most students did when they graduated. Therefore my case was not like some others. He had an idea, of course, that though I had collected and mounted birds, and knew their names and habits, I probably knew little about their anatomy. At any rate the first thing he did was to give me a badly mutilated old loon, from old alcohol, telling me to prepare the skeleton. This I did so well and so quickly that he expressed regret that he had not given me some better bird with unbroken bones. He gave me next a blue heron, but it being spring, I 'went collecting' in the vicinity, following my usual inclination, before breakfast and after laboratory hours, and brought in a number of incubated birds'-eggs. When Agassiz came into the laboratory, I was extracting and preserving the embryos, being interested in embryology. He at once exclaimed that he was delighted, and told me to put aside the skeletons and go right on with collecting and preparing embryo birds, and making drawings, etc. This I did all that season, obtaining about 2,000 embryos, mostly of sea birds, for he sent me to Grand Manan Island, etc., for that purpose. Before the end of the first year he gave me entire charge of the birds and mammals in the Museum, as well as the coral collection, which was large even then.

In the case of Hyatt, who went there just before I did, I think he was kept working over a lot of mixed fish skeletons, more or less broken, to 'see what he could make of them.' A little later he put Hyatt at work on the Unionidae, studying the anatomy as well as the shells. Within two years he put him on the Ammonites, a big collection having been received from Europe at that time. Hyatt, however, had never done anything in zoology or botany before he went to Agassiz and he found it hard to get a beginning, and so lost time. I mention these cases to show how different his methods were in different cases.

VI

HOW AGASSIZ TAUGHT PROFESSOR WILDER[1]

THE phrase adopted as the title of this article ['Louis Agassiz, Teacher'] begins his simple will. Agassiz was likewise an investigator, a director of research, and the founder of a great museum. He really was four men in one. Without detracting from the extent and value of the three other elements of his intense and composite American life— from his first course of lectures before the Lowell Institute in 1846 to the inauguration of the Anderson Summer School of Natural History at Penikese Island, July 8, 1873, and his address before the Massachusetts State Board of Agriculture, twelve days before his untimely death on December 14, 1873,—Agassiz was

[1] From an article by Professor Burt G. Wilder, of Cornell University, in *The Harvard Graduates' Magazine*, June, 1907. The extract is taken from a reprint with slight changes by the author, and is given with slight omissions by the present writer.

pre-eminently a teacher. · He taught his assistants; he taught the teachers in the public schools; he taught college students; he taught the public, and the common people heard him gladly. · His unparalleled achievements as an instructor are thus chronicled by his wife: ·

'A teacher in the widest sense, he sought and found his pupils in every class. But in America for the first time did he come into contact with the general mass of the people on this common ground, and it influenced strongly his final resolve to remain in this country. · Indeed the secret of his greatest power was to be found in the sympathetic, human side of his character. Out of his broad humanity grew the genial personal influence by which he awakened the enthusiasm of his audiences for unwonted themes, inspired his students to disinterested services like his own, delighted children in the school-room, and won the cordial interest, as well as the co-operation in the higher aims of science, of all classes, whether rich or poor.' ·

As a general statement the foregoing could not be improved. But the invitation to prepare this article contained a suggestion of particularity with which it is possible for me to comply.[2] The courses given by Agassiz on zoology and geology were attended by me during the three years (1859–62) of my pupilage with Jeffries Wyman, and the two years (1866–68) in which I was the assistant of Agassiz himself. Naturally, and also for special reasons, the deepest impression was made by the first and the last of these courses. With the former the charm of novelty intensified the great, indeed indescribable, charm of the speaker. No topic was to me so important as the general problem of animal life, and no expositor could compare with Agassiz. As an outlet for my enthusiasm each discourse was repeated, to the best of my ability, for the benefit of my companion, James Herbert Morse, '63, on the daily four-mile walk between Cambridge and our Brookline home. So sure was I that all the statements of Agassiz were correct and all his conclusions sound, that any doubts or

[2] Not only have I preserved all the letters from Agassiz, the first dated Sept. 4, 1866, and the last Nov. 25, 1873, but also my diaries in which are recorded all significant incidents and conversations from my first introduction in 1856 to the last interview, Sept. 5, 1873. [Note by Professor Wilder.]

criticisms upon the part of my acute and un-prejudiced friend shocked me as a reprehensible compound of heresy and lese-majesty.

The last course that I heard from Agassiz in Cambridge began on October 23, 1867, and closed on January 11, 1868. It was memorable for him and for me. At the outset he an-nounced that some progress had been made in the University toward the adoption of an elective system for the students, and that he proposed to apply the principle to his own instruction, and should devote the entire course of twenty-one lectures to the Selachians (sharks and rays), a group in which he had been deeply interested for many years, and upon which he was then preparing a volume. This limitation to a favorite topic inspired him to unusual energy and eloquence. My notes are quite full, but I now wish the lectures had been re-ported verbatim. This course was signalized also by two special innovations, viz.: the ex-hibition of living fish, and the free use of museum specimens. That, so far as possible, all biologic instruction should be objective was with Agassiz an educational dogma, and upon several notable occasions its validity had been demonstrated under very unfavorable conditions. Yet, during the five years of my attendance upon his lectures, they were seldom illustrated otherwise than by his ready and graphic blackboard drawings. The simple fact was that the intervals between his lectures were so crowded with multifarious, pressing, and never-ending demands upon his time and strength that he could seldom determine upon the precise subject long enough in advance for him, or any one else, to bring together the desirable specimens or even charts. The second lecture of the course already mentioned is characterized in my diary as 'splendid,' and as 'for the first time illustrated with many specimens.' At one of the later lectures, after speaking about fifteen minutes, he invited his hearers to examine living salmon embryos under his direction at one table, and living shark embryos under mine at another.

. Like those of Wyman, the courses given by Agassiz were Senior electives. I never heard of any examination upon them; nor is it easy to imagine Agassiz as preparing a syllabus, or formulating or correcting an examination-paper. His personality and the invariable attendance of teachers and other adults precluded the necessity of disciplinary measures. But his attitude toward student misconduct was clearly shown in an incident recorded by me elsewhere.[1] The method pursued by Agassiz with his laboratory students has been described by Scudder.[2] Although I was to prepare specimens at his personal expense, a somewhat similar test was applied. He placed before me a dozen young 'acanths' (dog-fish sharks), telling me to find out what I could about them. After three days he gave me other specimens, saying: 'When you go back to the little sharks you will know more about them than if you kept on with them now'—meaning, I suppose, that I should then have gained a better perspective.

Although, as I recall upon several occasions, Agassiz could express his views delightfully and impressively to a single auditor, his emi-

<hr>

[1] 'Agassiz at Penikese,' *American Naturalist*, March, 1898, p. 194. [Note by Professor Wilder.]

[2] See below, p. 40.

nently social nature and his lifelong habit rendered it easier for him to address a group of interested listeners. The following incident does not seem to have been recorded in my diary, but it is distinctly remembered. During the publication of the *Journey in Brazil*, a French translation was made by M. Félix Vogeli. With this the publishers desired to incorporate a chapter giving the latest views of Agassiz upon classification and evolution. In vain was he besought to write it. He hated writing, and was too busy. At last, in desperation, M. Vogeli came to the Museum with Mrs. Agassiz, and together they persuaded the Professor to dictate the required matter in the form of a lecture. For this, however, an audience was indispensable. The exigency was explained to the Museum staff; we assembled in the lecture-room, and the discourse began. To the dismay of some of us it proved to be in French, but we tried to look as if we comprehended it all.

Agassiz handled all specimens with the greatest care, and naturally had little patience with clumsiness; the following incident illustrates both his kindly spirit and his self-restraint. At one of the lectures he had handed down for inspection a very rare and costly fossil, from the coal-measures, I think; including the matrix, it had about the size and shape of the palm of the hand. He cautioned us not to drop it. When it had reached about the middle of the audience a crash was heard. The precious thing had been dropped by a new and somewhat uncouth assistant whom we will call Dr. X. He hastily gathered up the pieces and rushed out of the room. For a few seconds Agassiz stood as if himself petrified; then, without even an 'Excuse me,' he vanished by the same door. Presently he returned, flushed, gazing ruefully at the fragments in his hand, covered with mucilage or liquid glue. After a pause, during which those who knew him not awaited an explosive denunciation of gaucherie, Agassiz said quietly: 'In Natural History it is not enough to know how to study specimens; it is also necessary to know how to handle them' —and then proceeded with his lecture. ·

His helpful attitude toward prospective teachers was exhibited in the following incidents.

After my appointment to Cornell University in October, 1867, he arranged for me to give a course of six 'University Lectures,' and warned me to prepare for them carefully, because he should give me a 'raking down.' He attended them all (at what interruption of his own work I realize better now), and discussed them and my methods very frankly with me. Omitting the commendations, the following comments may be useful to other professorial tyros: 1. The main question or thesis should be stated clearly and concisely at the outset, without compelling the hearer to perform all the mental operations that have led the speaker to his own standpoint. 2. In dealing with the history of a subject, the value of each successive contribution should be estimated in the light of the knowledge at the period, not of that at the present time.

VII

HOW AGASSIZ TAUGHT PROFESSOR SCUDDER[1]

IT was more than fifteen years ago [from 1874] that I entered the laboratory of Professor Agassiz, and told him I had enrolled my name in the Scientific School as a student of natural history. He asked me a few questions about my object in coming, my antecedents generally, the mode in which I afterwards proposed to use the knowledge I might acquire, and, finally, whether I wished to study any special branch. To the latter I replied that, while I wished to be well grounded in all departments of zoology, I purposed to devote myself specially to insects.

'When do you wish to begin?' he asked.

'Now,' I replied.

This seemed to please him, and with an energetic 'Very well!' he reached from a shelf a huge jar of specimens in yellow alcohol.

[1] 'In the Laboratory with Agassiz,' by Samuel H. Scudder, from *Every Saturday* (April 4, 1874) 16, 369–370.

'Take this fish,' said he, 'and look at it; we call it a haemulon; by and by I will ask what you have seen.'

With that he left me, but in a moment returned with explicit instructions as to the care of the object entrusted to me.

'No man is fit to be a naturalist,' said he, 'who does not know how to take care of specimens.'

I was to keep the fish before me in a tin tray, and occasionally moisten the surface with alcohol from the jar, always taking care to replace the stopper tightly. Those were not the days of ground-glass stoppers and elegantly shaped exhibition jars; all the old students will recall the huge neckless glass bottles with their leaky, wax-besmeared corks, half eaten by insects, and begrimed with cellar dust. Entomology was a cleaner science than ichthyology, but the example of the Professor, who had unhesitatingly plunged to the bottom of the jar to produce the fish, was infectious; and though this alcohol had 'a very ancient and fishlike smell,' I really dared not show any aversion within these sacred precincts, and treated the alcohol as though it were pure water. Still I was conscious of a passing feeling of disappointment, for gazing at a fish did not commend itself to an ardent entomologist. My friends at home, too, were annoyed, when they discovered that no amount of eau-de-Cologne would drown the perfume which haunted me like a shadow.

In ten minutes I had seen all that could be seen in that fish, and started in search of the Professor—who had, however, left the Museum; and when I returned, after lingering over some of the odd animals stored in the upper apartment, my specimen was dry all over. I dashed the fluid over the fish as if to resuscitate the beast from a fainting-fit, and looked with anxiety for a return of the normal sloppy appearance. This little excitement over, nothing was to be done but to return to a steadfast gaze at my mute companion. Half an hour passed—an hour—another hour; the fish began to look loathsome. I turned it over and around; looked it in the face—ghastly; from behind, beneath, above, sideways, at a three-quarters' view—just as ghastly. I was in despair; at an early hour I concluded that lunch was necessary; so, with infinite relief, the fish was carefully replaced in the jar, and for an hour I was free. On my return, I learned that Professor Agassiz had been at the Museum, but had gone, and would not return for several hours. My fellow-students were too busy to be disturbed by continued conversation. Slowly I drew forth that hideous fish, and with a feeling of desperation again looked at it. I might not use a magnifying-glass; instruments of all kinds were interdicted. My two hands, my two eyes, and the fish: it seemed a most limited field. I pushed my finger down its throat to feel how sharp the teeth were. I began to count the scales in the different rows, until I was convinced that that was nonsense. ‹ At last a happy thought struck me—I would draw the fish; and now with surprise I began to discover new features in the creature. Just then the Professor returned.

'That is right,' said he; 'a pencil is one of the best of eyes. · I am glad to notice, too, that you keep your specimen wet, and your bottle corked.'

4

With these encouraging words, he added:
'Well, what is it like?'

He listened attentively to my brief rehearsal
of the structure of parts whose names were
still unknown to me: the fringed gill-arches
and movable operculum; the pores of the head,
fleshy lips and lidless eyes; the lateral line, the
spinous fins and forked tail; the compressed
and arched body. When I had finished, he
waited as if expecting more, and then, with an
air of disappointment:

'You have not looked very carefully; why,'
he continued more earnestly,' you haven't
even seen one of the most conspicuous features
of the animal, which is as plainly before your
eyes as the fish itself; look again, look again!'
and he left me to my misery.

I was piqued; I was mortified. Still more of
that wretched fish! But now I set myself to
my task with a will, and discovered one new
thing after another, until I saw how just the
Professor's criticism had been. The afternoon
passed quickly; and when, toward its close,
the Professor inquired:

'Do you see it yet?'

[44]

'No,' I replied, 'I am certain I do not, but I see how little I saw before.'

'That is next best,' said he, earnestly, 'but I won't hear you now; put away your fish and go home; perhaps you will be ready with a better answer in the morning. I will examine you before you look at the fish.'

This was disconcerting. Not only must I think of my fish all night, studying, without the object before me, what this unknown but most visible feature might be; but also, without reviewing my new discoveries, I must give an exact account of them the next day. I had a bad memory; so I walked home by Charles River in a distracted state, with my two perplexities.

The cordial greeting from the Professor the next morning was reassuring; here was a man who seemed to be quite as anxious as I that I should see for myself what he saw.

'Do you perhaps mean,' I asked, 'that the fish has symmetrical sides with paired organs?'

His thoroughly pleased 'Of course! of course!' repaid the wakeful hours of the previous night. After he had discoursed most happily and enthusiastically—as he always did—upon the importance of this point, I ventured to ask what I should do next.

'Oh, look at your fish!' he said, and left me again to my own devices. In a little more than an hour he returned, and heard my new catalogue.

'That is good, that is good!' he repeated; 'but that is not all; go on;' and so for three long days he placed that fish before my eyes, forbidding me to look at anything else, or to use any artificial aid. . 'Look, look, look,' was his repeated injunction.

This was the best entomological lesson I ever had—a lesson whose influence has extended to the details of every subsequent study; a legacy the Professor has left to me, as he has left it to many others, of inestimable value, which we could not buy, with which we cannot part. .

A year afterward, some of us were amusing ourselves with chalking outlandish beasts on the Museum blackboard. We drew prancing starfishes; frogs in mortal combat; hydra-headed worms; stately crawfishes, standing on their tails, bearing aloft umbrellas; and grotesque fishes with gaping mouths and staring eyes. The Professor came in shortly after, and was as amused as any at our experiments. He looked at the fishes.

'Haemulons, every one of them,' he said; 'Mr. —— drew them.'

True; and to this day, if I attempt a fish, I can draw nothing but haemulons.

The fourth day, a second fish of the same group was placed beside the first, and I was bidden to point out the resemblances and differences between the two; another and another followed, until the entire family lay before me, and a whole legion of jars covered the table and surrounding shelves; the odor had become a pleasant perfume; and even now, the sight of an old, six-inch, worm-eaten cork brings fragrant memories.

The whole group of haemulons was thus brought in review; and, whether engaged upon the dissection of the internal organs, the preparation and examination of the bony framework, or the description of the various parts, Agassiz's training in the method of observing facts and their orderly arrangement was ever accompanied by the urgent exhortation not to be content with them.

· 'Facts are stupid things,' he would say, 'until brought into connection with some general law.' ·

At the end of eight months, it was almost with reluctance that I left these friends and turned to insects; but what I had gained by this outside experience has been of greater value than years of later investigation in my favorite groups.[1]

[1] Professor Edward S. Morse writes: 'As I remember Mr. Scudder's article, . . . he has stated clearly the method of Agassiz's teaching—simply to let the student study intimately one object at a time. Day after day he would come to your table and ask you what you had learned, and thus keep you at it for a week. My first object put before me was a common clam, *Mya arenaria.*'

VIII

THE DEATH OF AGASSIZ—HIS
PERSONALITY[1]

IN later years the robust constitution and
herculean frame of Agassiz showed the
effects of his extraordinary and multifarious labors, for it must be confessed that he
was not careful of his bodily welfare. In the
year 1869 he suffered a temporary break-
down of a very threatening sort, and for months
was in seclusion, forbidden by his medical
advisers even to think. His own wise efforts,
and a quiet spring passed in the village of
Deerfield, Connecticut, brought about his recovery, so that three years of activity were

[1] The materials for this sketch are drawn from several
sources—chiefly the Life by Marcou (which I have used
with some caution) and the Life by Mrs. Agassiz. I had
wished to preserve the words of Marcou throughout (with
judicious omissions), but this wish was defeated by certain
persons who, for reasons unknown to me, have the power to
prevent the use of adequate quotations from him. I have
followed him where I had no other guide, and no ground for
suspecting him of bias. The composition, and to some
extent the interpretation of the facts, are my own.

yet to be vouchsafed him. But the strain of his lectures, of his correspondence, of his labors at and for the Museum, was perilous. On the second of December, 1873, he gave a lecture, his last, on 'The Structural Growth of Domestic Animals,' before the Massachusetts Board of Agriculture at Fitchburg. On the third he dined with friends; on the fifth he was present at a family gathering—and smoked cigars, defying the orders of his physician. But the end was not far off. He spoke of a dimness of sight; he complained of feeling 'strangely asleep.' On the morning of the sixth he went as usual to the Museum, but with a sense of great weariness he shortly returned to his room, where he lay down, never to depart from it alive. The disease was a paralysis of the organs of respiration, beginning with the larynx. · He had every care from his friends Dr. Brown-Séquard, who immediately came from New York, and Dr. Morrill Wyman; and the last few days of his life were passed, not in great suffering, with his loving family around him. Nothing, however, could arrest the progress of the malady.

Agassiz, it is said, had been afraid of softening of the brain, and of a long and painful illness like that which preceded the death of his friend Professor Bache; it had been his hope that he might rather go quickly. · Yet it was not easy for him to think of dying, when his imagination teemed with projects, and when the two main visions of his life were on the point of being fully accomplished, in the great Museum and the Anderson School of Natural History on the island of Penikese. Stricken though he was, he clung to life, nor did he give up all hope of recovery until the last day. Still there was a change of demeanor, for the aims of his career as a scientist were now less obtrusive in his mind than thoughts of his family. And with the arrival of Dr. Brown-Séquard he resumed the language of his youth, so that his last words were uttered in French. In the closing hours, when at length all hope was abandoned, he was more than once heard to say: '*Tout est fini.*' On the eighth day, when death itself was approaching, his family and friends—among these, Pourtalès—withdrew to an adjoining room, keeping watch over the

patient through the open door. While Pourtalès was standing there in his turn, not long
after ten o'clock at night, Agassiz lifted himself up in bed, and said with emphasis: '*Le jeu
est fini.*' Then, sinking back, he passed away.
'The play is done. *Plaudite.*'. For Agassiz
life was a game, full of motion, crowded with
incident. He could not understand the com-
plaint of those who found time hanging heavily
upon their hands, and who sought ways of
killing it. He, who had 'no time for making
money,' would gladly have borrowed an extra
life or two for study and teaching. From the
outset he had unwavering confidence in himself. He would be 'the first naturalist of his
time, a good citizen, a good son, beloved of
those who knew him.' He was not to follow
others; he would lead in his own path, which
should be the right path, and others should
follow him. ·

Agassiz was somewhat above the average in
height. His body was well formed, his shoul-
ders broad and square, his figure powerful,
firmly set upon rather small feet that served
him well in walking and climbing. With a

[52]

quick, elastic step, he was an excellent pedestrian, and quite at home in the mountains. As a boy he became proficient in swimming and in the management of boats. To bodily fear he was a stranger. His hands were large and shapely, and very skilful. Never a finished draughtsman, he was none the less expert in representing, with swift, sure strokes, the essential structure of the object he wished to recall or explain. He was deft, too, with the dissecting-knife and the microscope, and with the geologist's hammer. His neck (the weak part, as his fatal illness showed) was rather short; his head was fine and large. In later years his hair, of a chestnut color, deserted his brow, but he wore it full at the sides and back, and this, with the side-whiskers of the day, tended to conceal his ears. The head itself was admirable, the forehead high and broad, the chin shapely, the countenance frank and open. The mouth was wide, the lips full and smiling, the expression as a whole altogether amiable and intelligent. His aquiline nose, with well-developed nostrils, sharply set off by the oblique lines on either side, helped to give him an air of sagacity. But it was the magnificent, fascinating eyes, young, kindly, and searching, that above all gave life to that animated countenance. To those eyes nothing was commonplace.[1]

·Agassiz spoke French with a slight drawl characteristic of the section of Switzerland in which he was born. When he came to America in 1846, he rapidly acquired a command of English, and he eventually wrote and spoke the language with great facility, though his speech never ceased to betray his foreign origin.[2]

[1] Compare Clara Conant Gilson, 'Agassiz at Cambridge,' in *Frank Leslie's Popular Monthly*, December, 1891: 'He was a man of fine figure and striking appearance, not too much of the *embonpoint*, not too tall, but just tall enough to constitute a finely developed physique. His head was grand, of perfect intellectual shape, and commanded your admiration as you gazed. He was but slightly bald, his hair was of a beautiful brown, soft and fine, and fell lovingly over the collar of his coat. His face was of well-rounded contour, with a large, expressive mouth, and features indicative of great character and decision. His eyes were the feature of his face, *par excellence*. They were of a beautiful bright brown, full of tenderness, of meaning and earnestness—a liquid brown eye, that would moisten with tears of emotion as thoughts of his Creator came rushing to mind, while he traced His footsteps in the sciences he studied. His eyes mirrored his soul. I think there was never but one pair of eyes such as Professor Louis Agassiz's.'

[2] See Clara Conant Gilson, in the article just cited: He had

With his superabundant physical, mental, and emotional energy, he was a natural orator; he was fond of an audience, and gratified by applause. No one ever possessed a greater talent for making natural science popular; even when his discourse became highly technical, his auditors hung upon his words. His method of exposition was very clear and simple. He studiously avoided the error of dragging the listener through all the processes by which the speaker has arrived at a particular truth, and quickly came to the point. In lecturing, his personal magnetism counted for much; he readily communicated his enthusiasm to others.

He was easily moved to tears or to laughter. · In his earlier life he was seldom angry, or seldom showed it, but otherwise made no attempt to hide his feelings, being a perfect

a few striking peculiarities of pronunciation, one or two of which cling to me with great pertinacity even now. One, in particular, is fresh in my memory. For example, the words respiratory and perspiratory he would accent on the third syllable—*rat;* and, bless me, if to this day I don't have to think twice before I am sure which is right! This shows what indelible impressions his words left upon his pupils.

[55]

child of nature. Later he became less demonstrative, save when he was angry. In the last twenty years of his life he not infrequently lost his temper, though he would not utterly forget what he was saying; and, however heated the discussion might become, he never ceased to be a gentleman. Neither indecency nor aught approaching thereto ever issued from his lips. As a youth in Switzerland, during his life as a student, and even when he was a teacher at Neuchâtel, he was fond of singing, and he liked to yodel after the fashion of the Swiss and Tyrolese mountaineers, but he gave this up when he came to America.

. Here his recreations were mostly social. He was the friend of Longfellow, Lowell, and Whittier; he was the friend of laborers and fishermen. In society he liked to encounter men of wealth and influence, for he had by nature, and also learned from Alexander von Humboldt, some of the arts of the courtier. 'It would be difficult,' says Dr. Charles D. Walcott,[1] 'to measure his influence in the way of causing men of political and commercial

[1] *Smithsonian Miscellaneous Collections* 50. 217 (1908).

power to realize that the support of scientific research, and the diffusion of knowledge thereby gained, depend largely on them.' In other natural scientists he was prone to discover too much self-satisfaction, and too much personal curiosity, against which he hardly knew how to protect himself. But with the group of younger scientists he himself developed, though now and then one or another grew mutinous, he was, during most of the time, on the best of terms. His own early schooling in the classics gave him a relish for scholars, and he was pleased with the company of historians and lawyers. For military men he did not care, but he liked naval officers and sea-captains. He paid little attention to matters of dress, certainly as regards his own person. He was gratified by the marks of distinction conferred upon him at home and abroad, but took little subsequent thought of the ribbons, badges, and diplomas, keeping them, but not very carefully, and never making a parade of them. ·

Eloquent as a lecturer, he was also brilliant and persuasive in conversation, being, in appearance at least, quite unreserved, and open in his attempt to capture the good will of his auditor. However, if there was no covert artifice, there was at all events the native shrewdness of the Swiss peasant to reckon with, and doubtless the subtlety of genius—which will not, or cannot, always reveal itself in full. In his later years, accordingly, though his winning manners and his desire that you should completely display your thought to him might lead you to suppose him utterly open with you, you might in the end discover that you had not fathomed his soul, that there was that in him which could not be taken captive, and that there might be a silent invincible rejection on his part of something within you which was foreign to him.

. In Agassiz the theoretical and the practical life were well balanced. He was both a visionary and a man capable of bringing his visions to pass. No philosophical conception was too general for him, and no detail of observation or inference too small. No fact could appear too slight for his intense and comprehensive scrutiny, and his memory for minute resemblances and differences was vast; yet the enduring

[58]

quality of his work arose from his sense of order, and from the soundness and rigor of his principles. He possessed not only physical, but intellectual and moral courage. In the face of hardship or difficulty he was undaunted, ever energetic at the moment, ever hoping for better times. His power of working was enormous, for he made virtually no false motions, but proceeded silently, swiftly, with no apparent effort, and for long periods without interruption. ·

Much has been said by his friends of the depth and sincerity of his sentiments in point of religion. But he had little sympathy with clergymen, or with the definite forms in which the religious experience of man has expressed itself—though these forms are in their essence and development not unlike the natural forms which he so reverently studied. · One who knew him well affirms that in early manhood Agassiz, if not precisely a materialist, was at all events a sceptic; but his later studies, with mature reflection, led him to believe in a Divine Creator. The external universe became to him the language in which the Divine Being

5   [59]

conveys his ideas to man, and natural history
the discipline by which men interpret that
language. · Thus he says, in the *Essay on
Classification:* 'To me it appears indisputable
that this order and arrangement of our studies
are based upon the natural, primitive relations
of animal life—those systems, to which we have
given the names of the great leaders of our
science who first proposed them, being in
truth but translations into human language of
the thoughts of the Creator.   And if this is indeed so, do we not find in this adaptability of
the human intellect to the facts of creation,
by which we become instinctively, and, as I
have said, unconsciously, the translators of the
thoughts of God, the most conclusive proof of
our affinity with the Divine mind? And is
not this intellectual and spiritual connection
with the Almighty worthy of our deepest con-
sideration? If there is any truth in the belief
that man is made in the image of God, it is
surely not amiss for the philosopher to endeavor,
by the study of his own mental operations, to
approximate the workings of the Divine Reason,
learning from the nature of his own mind better

to understand the Infinite Intellect from which it is derived. Such a suggestion may, at first sight, appear irreverent. But who is the truly humble? He who, penetrating into the secrets of creation, arranges them under a formula, which he proudly calls his scientific system? or he who in the same pursuit recognizes his glorious affinity with the Creator, and in deepest gratitude for so sublime a birthright strives to be the faithful interpreter of that Divine Intellect with whom he is permitted, nay, with whom he is intended, according to the laws of his being, to enter into communion?'[1] Herein we may discern the secret of his power as a teacher.

. 'Agassiz's influence on methods of teaching in our community,' said Professor James, 'was prompt and decisive—all the more that it struck people's imagination by its very excess. The good old way of committing printed abstractions to memory never seems to have received such a shock as it encountered at his hands. · There is probably no public school teacher now [1896] in New England who will

[1] *Essay on Classification* (1859), pp. 9–10.

[61]

not tell you how Agassiz used to lock a student
up in a room full of turtle-shells, or lobster-
shells, or oyster-shells, without a book or a
word to help him, and not let him out till he
had discovered all the truths which the objects
contained. Some found the truths after weeks
and months of lonely sorrow; others never
found them. Those who found them were
already made into naturalists thereby—the fail-
ures were blotted from the book of honor and
of life. · "Go to nature; take the facts into
your own hands; look, and see for yourself!"—
these were the maxims which Agassiz preached
wherever he went, and their effect upon pedagogy was electric. ·. . . While on the Thayer
expedition [to Brazil, in 1865], I remember
that I often put questions to him about the
facts of our new tropical habitat, but I doubt
if he ever answered one of these questions of
mine outright. He always said: "There, you
see you have a definite problem. Go and look,
and find the answer for yourself." [1]

[1] William James, *Louis Agassiz, Words Spoken* . . . *at
the Reception of the American Society of Naturalists* . . .
[Dec. 30, 1896]. Pp. 9, 10. Cambridge, 1897.

IX

OBITER DICTA BY AGASSIZ[1]

NEVER try to teach what you yourself do not know, and know well. · If your school board insists on your teaching anything and everything, decline firmly to do it. It is an imposition alike on pupils and teacher to teach that which he does not know. Those teachers who are strong enough should squarely refuse to do such work. This much-needed reform is already beginning in our colleges, and I hope it will continue. It is a relic of mediaeval times, this idea of professing everything. When teachers begin to decline work which they cannot do well, improvements begin to come in. If one will be a successful teacher, he must

[1] The first nine of these utterances were taken down by Dr. David Starr Jordan at Penikese, in the summer of 1873, from Agassiz's talks to teachers; see *Popular Science Monthly* 40. 726–727, and Holder, *Louis Agassiz, his Life and Works*, 1893, pp. 173–176. The next five come from the article entitled 'Louis Agassiz, Teacher,' by Professor Burt G. Wilder, in *The Harvard Graduate's Magazine*, June, 1907, and the last three from Agassiz's posthumous article, 'Evolution and Permanence of Type,' in the *Atlantic Monthly*, Jan., 1874 (vol. 33).

firmly refuse work which he cannot do successfully.

· It is a false idea to suppose that everybody is competent to learn or to teach everything. Would our great artists have succeeded equally well in Greek or calculus? A smattering of everything is worth little. It is a fallacy to suppose that an encyclopaedic knowledge is desirable. The mind is made strong, not through much learning, but by the thorough possession of something. .

Lay aside all conceit. Learn to read the book of nature for yourself. Those who have succeeded best have followed for years some slim thread which has once in a while broadened out and disclosed some treasure worth a life-long search.

A man cannot be a professor of zoology on one day, and of chemistry on the next, and do good work in both. As in a concert all are musicians—one plays one instrument, and one another, but none all in perfection.

You cannot do without one specialty; you must have some base-line to measure the work and attainments of others. For a general view of the subject, study the history of the sciences. · Broad knowledge of all nature has been the possession of no naturalist except Humboldt, and general relations constituted his specialty. ·

Select such subjects that your pupils cannot walk without seeing them. Train your pupils to be observers, and have them provided with the specimens about which you speak. If you can find nothing better, take a house-fly or a cricket, and let each hold a specimen and examine it as you talk.

In 1847 I gave an address at Newton, Massachusetts, before a Teachers' Institute conducted by Horace Mann. My subject was grasshoppers. I passed around a large jar of these insects, and made every teacher take one and hold it while I was speaking. If any one dropped the insect, I stopped till he picked it up. This was at that time a great innovation, and excited much laughter and derision. There can be no true progress in the teaching of natural science until such methods become general.

There is no part of the country where, in the summer, you cannot get a sufficient supply of the best specimens. Teach your children to bring them in themselves. Take your text from the brooks, not from the book-sellers. It is better to have a few forms well known than to teach a little about many hundred species. Better a dozen specimens thoroughly studied as the result of the first year's work, than to have two thousand dollars' worth of shells and corals bought from a curiosity-shop. The dozen animals would be your own.

The study of nature is an intercourse with the highest mind. You should never trifle with nature. At the lowest her works are the works of the highest powers—the highest something, in whatever way we may look at it.

It is much more important for a naturalist to understand the structure of a few animals than to command the whole field of scientific nomenclature.

Methods may determine the result.

The only true scientific system must be one

in which the thought, the intellectual structure, rises out of, and is based upon, facts.

He is lost, as an observer, who believes that he can, with impunity, affirm that for which he can adduce no evidence.

Have the courage to say: 'I do not know.' ·

· Since the ability of combining facts is a much rarer gift than that of discerning them, many students lost sight of the unity of structural design in the multiplicity of structural detail.[1]

· It cannot be too soon understood that science is one, and that whether we investigate language, philosophy, theology, history, or physics, we are dealing with the same problem, culminating in the knowledge of ourselves. Speech is known only in connection with the organs of man, thought in connection with his brain, religion as the expression of his aspirations, history as the record of his deeds, and physical sciences as the laws under which he lives.[2] ·

The most advanced Darwinians seem reluctant to acknowledge the intervention of an

[1] *Atlantic Monthly* 33. 93.
[2] *Atlantic Monthly* 33. 95.

intellectual power in the diversity which obtains
in nature, under the plea that such an admission implies distinct creative acts for every
species. What of it, if it were true? Have
those who object to repeated acts of creation
ever considered that no progress can be made
in knowledge without repeated acts of thinking?
And what are thoughts but specific acts of
the mind? Why should it then be unscientific
to infer that the facts of nature are the result
of a similar process, since there is no evidence
of any other cause? The world has arisen in
some way or other. How it originated is the
great question, and Darwin's theory, like all
other attempts to explain the origin of life, is
thus far merely conjectural. I believe he has
not even made the best conjecture possible in
the present state of our knowledge.

· The more I look at the great complex of the
animal world, the more sure do I feel that we
have not yet reached its hidden meaning, and
the more do I regret that the young and ardent
spirits of our day give themselves to speculation
rather than to close and accurate investigation.[3]

[3] *Atlantic Monthly* 33. 101.

X

PASSAGES FOR COMPARISON WITH THE METHOD OF AGASSIZ

BOECKH ON THE STUDY OF HISTORY AND LITERATURE[1]

THE person who first seeks to acquire a general survey of a science, and then gradually to descend to details, will never attain to sound and exact knowledge, but will for ever dissipate his energies, and, knowing many things, will yet know nothing. In his lectures on the Method of Academical Study, Schelling remarks with great justice that, in history, to begin with a survey of the entire past is in the highest degree useless and injurious, since it gives one mere compartments for knowledge, without anything to fill them. In history, his advice is, first study one period in detail, and from this broaden out in all

[1] August Boeckh, *Encyclopädie und Methodologie der Philologischen Wissenschaften*, pp. 46–47.

[69]

directions. · For the study of language and literature (which corresponds with history in its most general sense) a similar procedure is the only right one. · Everything in science is related; although science itself is endless, yet the whole system is pervaded with sympathies and correspondences. Let the student place himself where he will—so long as he selects something significant and worth while,—and he will be compelled to broaden out from this point of departure in every direction in order to reach a complete understanding of his subject. From each and every detail one is driven to consider the whole; the only thing that matters is that one go to work in the right way, with strength, intelligence, and avidity. Let one choose several different points of departure, working through from each of them to the whole, and one will grasp the whole all the more surely, and comprehend the wealth of detail all the more fully. Accordingly, by sinking deep into the particular, one most easily avoids the danger of becoming narrow.

PASSAGES FOR COMPARISON

FROM THE SYMPOSIUM OF PLATO

· [The passage is thus summarized by Jowett: 'He who would be truly initiated should pass from the concrete to the abstract, from the individual to the universal, from the universal to the universe of truth and beauty.'] ·

· *Diotima.* . . . These are the lesser mysteries of love, into which even you, Socrates, may enter; ·to the greater and more hidden ones which are the crown of these, and to which, if you pursue them in a right spirit, they will lead, I know not whether you will be able to attain. But I will do my utmost to inform you, and do you follow if you can. · He who would proceed aright in this matter should begin in youth to visit beautiful forms; and first, if he be guided by his instructor aright, to love one such form only—out of that he should create fair thoughts. And soon he will of himself perceive that the beauty of one form is akin to the beauty of another; and then, if beauty of form in general is his pursuit, how foolish would

[1] Plato, *Symposium. The Dialogues of Plato*, translated by Jowett, New York, Oxford University Press, 1892, 1. 580–582.

[71]

he be not to recognize that the beauty in every form is one and the same! And when he perceives this, he will abate his violent love of the one, which he will despise and deem a small thing, and will become a lover of all beautiful forms. In the next stage he will consider that the beauty of the mind is more honorable than the beauty of the outward form. So that if a virtuous soul have but a little comeliness, he will be content to love and tend him, and will search out and bring to the birth thoughts which may improve the young, until he is compelled to contemplate and see the beauty of institutions and laws, and to understand that the beauty of them all is of one family, and that personal beauty is a trifle. And after laws and institutions he will go on to the sciences, that he may see their beauty, being not, like a servant, in love with the beauty of one youth or man or institution, himself a slave, mean and narrow-minded, but drawing towards and contemplating the vast sea of beauty, he will create many fair and noble thoughts and notions in boundless love of wisdom; until on that shore he grows and waxes strong, and at last the vision is revealed to him of a single science, which is the science of beauty everywhere. . . .

He who has been instructed thus far in the things of love, and who has learned to see the beautiful in due order and succession, when he comes toward the end will suddenly perceive a nature of wondrous beauty (and this, Socrates, is the final cause of all our former toils)—a nature which in the first place is everlasting, not growing and decaying, or waxing and waning; secondly, not fair in one point of view and foul in another, or at one time or in one relation or at one place fair, at another time or in another relation or at another place foul, as if fair to some and foul to others, or in the likeness of a face or hands or any other part of the bodily frame, or in any form of speech or knowledge, or existing in any other being, as, for example, in an animal, or in heaven, or in earth, or in any other place; but beauty absolute, separate, simple, and everlasting, which without diminution and without increase, or any change, is imparted to the ever-growing and perishing beauties of all other things.

He who from these ascending under the influence of true love, begins to perceive that
beauty, is not far from the end.   And the true
order of going, or being led by another, to the
things of love is to begin from the beauties of
earth, and mount upwards for the sake of that
other beauty, using these as steps only, and
from one going on to two, and from two to
all fair forms, and from fair forms to fair
practices, and from fair practices to fair notions,
until from fair notions he arrives at the notion
of absolute beauty, and at last knows what
the essence of beauty is.

Milton Keynes UK
Ingram Content Group UK Ltd.
UKHW011930021123
431764UK00004B/384